BRADY

POCKET REFERENCE FOR CHEMICAL, BIOLOGICAL, AND RADIOLOGIC AGENTS

Lt. John P. Bastin, MHS PA-C
Head of Emergency Medicinal Services and Disaster Preparedness
Branch Medical Clinic, NAS Brunswick, ME

PEARSON
Prentice Hall

Upper Saddle River, NJ 07458

A Cataloging-in-Publication record for this book is available from the Library of Congress.

Brady's Pocket Reference for Chemical, Biological, and Radiologic Agents was designed to provide you, the EMT-B or First Responder, with a handy, pocket-size guide to which you can quickly refer.

Upper Saddle River, NJ 07458

The information in this book expresses the views of the author and not those of the United States Navy.

Printed in the United States of America

10 9 8 7 6 5 4 3 2 1

ISBN 0-13-119001-6

PEARSON

Prentice Hall

This guide is dedicated to all U.S. Navy Hospital Corpsmen, past, present, and future, who unselfishly answer the call to duty throughout the United States and the world.

It is also dedicated to all civilian emergency responders, who are the first line of defense in preserving the lives of those they serve.

What you do on a daily basis can be summed up in the ending of the U.S. Air Force Pararescue Creed:

"These things I do so others may live."

Contents

Rick Hoffman, EMT-P, WMD Technician, HAZ-MAT specialist.

Sandy Waggoner, EMT-P/FF/EMSI, EHOVE Ghrist Adult Career Center.

Chris Hafley, EHOVE Adult Career Center.

Captain Robert C. Krause, Toledo Fire and Rescue, Toledo, Ohio.

Scott Krager, Emergency Response Coordinator for American Airlines, Capt. Tiawah Fire Dept.

This guide was written following an extensive review of the most current chemical, biological, and radiologic references and is a compilation of relevant information derived from that review. This reference was initially developed for and was used by U.S. Navy Medical Department personnel deployed in support of Operation Iraqi Freedom.

The purpose of this guide is to provide all First Responders, as well as hospital-based medical professionals, with a concise but complete reference that is relevant to a weapons-of-mass-destruction (WMD) incident. Category A and B biological agents that are designated by the Centers for Disease Control (CDC) as having the greatest potential for use as biological weapons are reviewed in this guide.

Specific information regarding personnel protective equipment (PPE) is not addressed in this guide. The use and selection of appropriate PPE is incident and agency specific, and First Responders are encouraged to refer to their agencies' protocols for specific guidance regarding the use of PPE.

This reference is not intended to be used as a single WMD reference, and First Responders should use this guide in conjunction with their departments' specific standard operating guidelines as well as their own experience and training.

➤ IMMEDIATE (RED)

Patients who have a reasonable chance of surviving if life-threatening conditions are corrected and they are rapidly transported to a definitive care facility. If treatment will require extensive time and effort on the part of the rescuers and be a detriment to other patients, triage the patient as expectant.

Example: Severe bleeding
Tension pneumothorax

➤ DELAYED (YELLOW)

Patients who will survive for a reasonable amount of time after receiving initial first aid. These patients may require extensive treatment at a later time.

Example: Femur fractures

➤ MINIMAL (GREEN)

Minimally injured patients, or "walking wounded." Treat these patients with basic first aid.

Example: Lacerations without hemorrhage
Minor burns

➤ EXPECTANT/DECEASED (BLACK)

Already deceased or critically injured patients who have very little chance of survival or who would only survive with extensive resuscitative efforts.

Example: Extensive burns
Extensive amputations

"START" TRIAGE

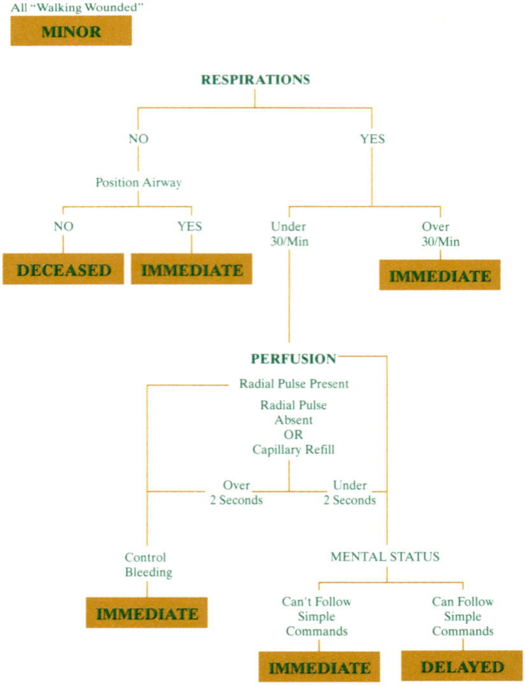

Mass victim triage is a method of triage that is based on the correlation between the severity of the patient's injuries and the motor component of the Glasgow Coma Scale (GCS).

This system relies on the patient's ability or inability to respond to verbal commands and demonstrate purposeful extremity movement as an indicator of the severity of his injuries.

Perform mass victim triage by first announcing,

"Everyone who can hear me, come toward my voice."

Initially triage these patients as minimal.

Next, announce,

"Everyone who can hear me but is unable to come toward my voice, raise an arm or leg."

Initially triage these patients as delayed.

Quickly assess the remaining patients.

Triage patients as immediate if they have spontaneous respirations and a palpable pulse or if they have spontaneous respirations after you have opened the airway.

If there are no spontaneous respirations and no palpable pulse, triage the patient as expectant.

Re-evaluate and retriage patients as they arrive at the staging area. Change triage categories if the patient's condition changes.

Sort contaminated patients into two separate triage categories:

➤ AMBULATORY PATIENTS

Patients who are able to understand directions, talk, and walk unassisted.

Triage most of them as minimal (Green).

➤ NONAMBULATORY PATIENTS

Patients who are unconscious, unresponsive, or unable to move without assistance.

Ambulatory patients should undergo decontamination in the following order:

1. Patients who were closest to the point of release and who report that they were exposed to an aerosol or vapor.
2. Patients who have some evidence of liquid agent on their skin or clothes and are experiencing severe signs and symptoms (shortness of breath, chest tightness).
3. Patients who were not close to the point of release but have evidence of the agent on their clothes or skin and are symptomatic.
4. Patients suffering from conventional wounds, especially open wounds.
5. The lowest decontamination priority is for ambulatory patients who were far away from the point of release and are asymptomatic: the "walking worried."

Triage nonambulatory patients prior to decontamination using standard medical triage systems (i.e., NATO, START, mass casualty triage).

1. The first priority is for those patients who are medically triaged as immediate and who are in need of immediate life-saving procedures that can be done quickly with the medical resources available on site. Patients with signs of severe to moderate nerve-agent exposure may be the highest priority for decontamination within this category.
2. The next priority is for those patients medically triaged as delayed. These patients may have serious injuries and require definitive care but can wait for a short period of time before receiving treatment.
3. Triage as delayed those patients with no known or suspected exposure.
4. The lowest priority triage category for decontamination contains those patients medically triaged as expectant.

Make attempts to correct all life-threatening injuries prior to proceeding with decontamination.

Rapid physical removal of an agent is the single most important action associated with decontamination. Physical removal includes:

- disrobing
- blotting off any agent that is visible on the skin
- using absorbents to soak up the agent
- flushing or showering with large quantities of water

Removing the patient's clothes will greatly reduce exposure to the agent (60–90 percent reduction).

- Chemical agents may persist in the patient's clothes, and the "off gassing" of the vapor may be harmful to rescuers and ED personnel.
- Removing the patient's clothes is usually all that is needed for decontamination after an exposure to a biological agent or vapor.

Water alone is an excellent decontamination solution. "The solution to pollution is dilution."

- Water alone will cause only a marginal breakdown of chemical agents.
- Soaps will aid in dissolving oily substances such as mustard agents.
- Liquid soaps are quicker than solid soaps and have less potential for abrading the skin (abrading the skin may result in increased absorption of the agent).
- Skin should be gently but thoroughly scrubbed with sponges.

Blot the skin to remove chemical or radiological agents. Wiping will spread the agent and increase the area of exposure on the skin.

Ensure that you are wearing appropriate personnel protective equipment (PPE) before beginning decontamination of patients.

Ensure that decontamination corridors are properly marked and that there is clear flow of patients from the "hot zone" or dirty side through the decontamination process into the "cold zone" or clean side.

Contaminated patients should be separated from uncontaminated patients.

Quickly attempt to correct all life-threatening injuries and conditions prior to beginning patient decontamination.

Attempt to segregate men from women during decontamination, and keep small children with their parents.

Wetting down the patients as they start to disrobe will speed up the decontamination process and is recommended when decontaminating biological or radiological exposures.

Encourage patients to remove as much clothing as possible, at least down to their undergarments, prior to showering.

Encourage patients with pre-existing medical conditions, young children, and the elderly to enter the stream of water slowly to minimize cold shock.

Deliver water for decontamination at high volumes and water pressures at a minimum of 60 psi to adequately remove the agent (household water pressure is between 60–90 psi).

High-pressure water could actually force the agent through the patient's clothes and onto his skin, increasing the exposure.

The length of showering time is incident specific but may be as long as 3–5 minutes under ideal circumstances.

➤ COLD-WEATHER DECONTAMINATION

The breakpoint temperature for outdoor decontamination is 65°F/18.3°C.

At ambient air temperatures below 65°F/18.3°C, patients will be much less willing to participate in outdoor showering because of the perception of significant discomfort.

Encouraging patients to enter the stream of water gradually to adjust to the cold can minimize the risk of cold shock.

Allow small children to remain with parents, and encourage parents to assist with the decontamination of the child.

Guidelines for Cold-Weather Decontamination

- *Ambient air temperatures above 65°F/18.3°C*
  Outdoor water decontamination without restrictions.

- *Ambient air temperatures above 35°F/1.7°C*
  Outdoor patient collection, assessment, and decontamination are acceptable, but move patients to a heated enclosure after decontamination.
- *Ambient air temperatures less than 35°F/1.7°C*
  Carry out collection and assessment of patients either outdoors or indoors, but carry out decontamination indoors.

Severe Weather Conditions

Use dry decontamination when severe weather conditions exist, that is, when the ambient air temperature is less than 35°F/1.7°C or when other methods of decontamination are not available.

Dry-Decontamination Techniques

Dry decontamination is used when water is not available or when the use of water is not practical.

- Use any absorbent material to remove the agent, for example, paper towels, sand, flour, or oil absorbent.

Removing the patient's clothing will reduce exposure to an agent by 60–80 percent.

- Blot the patient's skin to remove any visible agent.

After dry decontamination is completed, consider some other form of wet decontamination:

- an indoor swimming pool
- indoor facilities with showers
- water mist from opposing fire hose nozzles

Use dry decontamination for patients whose bodies may not be able to tolerate a cold shock:

- patients with pre-existing medical conditions
- the very young and elderly
- patients with concomitant trauma, burns, and/or shock

Since getting wet gradually might lessen the cold shock, encourage patients to do so if they have pre-existing medical conditions or are very young or old.

Trauma patients, depending on the severity of their injuries and degree of shock, will most likely not be able to tolerate a decontamination-induced cold shock.

Post-Decontamination Procedures

Prevent unnecessary exposure of patients to the environment during cold weather.

When possible, provide some form of shelter, dry clothing, or blankets.

If possible, separate men and women, as this may minimize some of the psychological stress.

Observe the patients for signs of shivering. Shivering generates body heat and is a ***normal*** **body response to a cold environment**.

If shivering stops, hypothermia is imminent, and medical personnel should immediately evaluate the patients.

Continually monitor patients not only for hypothermia but also for delayed reactions to the agent or for occult signs of trauma.

Covering the patient's head and upper body with a space blanket after decontamination is an ideal method of correcting for hypothermia and preventing further heat loss.

Section II: Biological Agents

Encapsulated spore-forming organism.

A large number of spores are required to cause disease in humans, 8,000–10,000.

Two toxins associated with the organism's capsule are required to cause symptoms:

- edema toxin: swelling
- lethal toxin: bleeding

If anthrax is suspected, notify the laboratory prior to submitting blood cultures, because a special culture media must be used. A contaminate resembling *Bacillus cereus* will grow, causing the specimen to be discarded if the blood culture is not planted on sheep-blood agar.

You will see colonies of large gram-positive rods in 6–8 hours on sheep-blood agar.

Obtain blood cultures prior to initiating treatment.

➤ CUTANEOUS ANTHRAX

Incubation period 1–7 days.

The lesion will follow a typical progression.

The lesion begins as a painless papule (may itch) greater in size than a hemorrhagic vesicle, with central necrosis evolving into a black escher.

Antibiotics will not stop lesion progression but may lessen constitutional symptoms.

The lesion aspirate will grow anthrax organisms if the patient is not already on antibiotics.

The organism is found around the edges of lesion; therefore, swab or biopsy around the edge of the lesion when obtaining a specimen for culture.

➤ OROPHARYNGEAL ANTHRAX

Can develop after the ingestion of contaminated meat. Incubation period 1–6 days.

Symptoms include swollen "bull" neck and enlarged cervical nodes.

The tan–white pseudomembranous oral lesions are easily confused with diphtheria.

Death is usually due to airway edema.

➤ INHALATIONAL ANTHRAX

An incubation period of up to 60 days after inhaling spores.

A large number of spores will germinate quickly and cause a rapid onset of disease.

A small number of spores will be ingested by macrophages and cause a slow onset of disease.

➤ NONSPECIFIC SYMPTOMS

Onset 2–5 days after exposure.

It may be confused with "influenza," but symptoms are not typical of influenza.

- Drenching continuous sweats
- Temperatures >104°F
- Mental status changes

- Dry nonproductive cough
- No rhinorrhea or congestion

The patient will appear sick, but the lung exam and complete blood count (CBC) will be normal.

➤ TERMINAL PHASE

The patient will deteriorate quickly.

You will occasionally see chest-wall edema.

Large pleural effusions will develop.

The resultant infection will travel to meninges early, causing hemorrhagic meningitis.

You will see mediastinum widening and clear lungs on a chest x-ray.

➤ DECONTAMINATION

Standard precautions for healthcare workers.

Remove all clothing, shoes, and jewelry.

Wash with copious amounts of water and mild liquid detergent.

Spores can remain on clothes, leading to possible secondary transmission.

➤ ANTHRAX TREATMENT

Postexposure Prophylaxis

Adults (including pregnant patients):
Ciproflaxin 500 mg BID for 60 days.

Alternate: Doxycycline 100 mg BID for 60 days.

Pediatric patients:
Ciproflaxin 20–30 mg/kg/day Q12 hour × 60 days.

Alternate: Doxycycline >8 years and >45 kg, 100 mg PO BID;
>8 years and <45 kg, 2.2 mg/kg PO BID;
<8 years, 2.2 mg/kg PO BID.
All dosages are for 60 days.

Cutaneous Anthrax Treatment

Adults (including pregnant patients):
Ciproflaxin 500 mg PO BID for 60 days.

Alternate: Doxycycline 100 mg PO BID for 60 days.

Pediatric patients:
Ciproflaxin 20–30 mg/kg/day Q12 hours × 60 days.
Maximum dosage 1 g/day.

Alternate: Same as for postexposure prophylaxis.

Inhalational, Oropharyngeal, and GI Anthrax Treatment

Adults (including pregnant patients):
Ciproflaxin 400 mg IV Q12 hours
or
Doxycycline 100 mg IV Q12 hours
with
Clindamycin 900 mg IV Q8 hours
and
Rifampin 300 mg IV Q8 hours.

Switch to PO when you are able to, and lower the dose to

Ciproflaxin 500 mg PO BID/Clindamycin 450 mg PO Q8 hour
and
Rifampin 300 mg PO BID for 60 days.

Inhalational/Oropharyngeal/GI Treatment

Pediatric patients:
Ciproflaxin 10–15 mg/kg IV Q12 hour
or
Clindamycin 7.5 mg/kg IV Q6 hour
and
Rifampin 20 mg/kg IV QD
× 60 days.

➤ DNA VIRUS

The infectious dose is unknown, although the disease is thought to be caused by only a few virons.

The infection occurs from close face-to-face contact.

Natural infection occurs following implantation of the virus on the oropharyngeal or respiratory mucosa.

An asymptomatic viremia develops 3–4 days after the virus migrates and replicates in regional lymph nodes. The virus then multiplies in the spleen, lymph nodes, and bone marrow.

Secondary viremia develops on about day 8 and is associated with fever and toxemia.

➤ INCUBATION PERIOD

12–14 days (average 7–17 days).

➤ NONSPECIFIC SYMPTOMS

Develop at the end of the incubation period:

- high fever
- malaise
- back pain
- prostration
- headache
- abdominal pain and delirium are occasionally present

➤ RASH

The nonspecific symptoms are followed by a maculopapular rash that appears on the oral and pharyngeal mucosa and on the face and forearms. It spreads to the trunk and legs.

The oral lesions ulcerate quickly, releasing large amounts of virus into the saliva.

Patients are contagious at the onset of oral rash.

Becomes vesicular within 1–2 days, later pustular.

Scabs form on day 8 or 9.

The rash is centrifugal in distribution (greater distribution of lesions on face and extremities). The lesions all evolve at the same rate, and are in the same stage of development.

You will see the rash on the palms of hands and the soles of feet.

➤ HEMORRHAGIC SMALLPOX

Uniformly fatal.

Pregnant women are more susceptible.

A shorter incubation period than nonhemorrhagic smallpox and more severe prodrome that is followed by:

- dusky erythema
- petechia
- frank hemorrhages into the skin and mucous membranes

Death usually occurs on the fifth or sixth day after the onset of the rash.

➤ MALIGNANT SMALLPOX

Abrupt onset of severe constitutional symptoms.

Confluent lesions develop slowly but do not progress to a pustular stage.

The lesions remain soft and velvety to the touch.

Skin appears fine grained like reddish-colored crepe rubber, occasionally with hemorrhages.

If the patient survives, the lesions disappear without forming scabs.

In severe cases, large amounts of the epidermis can peel away.

➤ VARIOLA MINOR

Symptoms are less severe than in the other forms of smallpox.

Sparse rash.

This milder form of the disease can be seen in patients with residual immunity from previous vaccination.

➤ LABORATORY CONFIRMATION

Collect specimens for laboratory confirmation only if you have been recently vaccinated (even if only on that day), or use personnel protective equipment (PPE).

To obtain vesicular fluid,

- de-roof lesions with a scalpel blade and
- collect the fluid on a cotton swab.

Pick off scabs with a pair of forceps.

Place specimens in a vacutainer tube.

Seal the junction of the rubber stopper and tube with tape.

Place the tube in a durable watertight container.

Contact the state health department laboratories for shipping instructions.

➤ POSTEXPOSURE TREATMENT

Vaccination within 3 days after exposure may prevent serious disease.

Vaccination 4–7 days after exposure offers partial protection.

Because contacts, even if infected, are not contagious until the onset of rash, check the temperature of all contacts in the evenings. Any increase in the temperature beyond 101°F during the 17-day period following the last exposure would suggest possible infection with smallpox.

Cidofovir has been proposed to treat smallpox.

Consider VIG if contraindications to vaccination exist.

➤ VACCINIA IMMUNE GLOBULIN (VIG)

Indicated to treat vaccine complications.

- Progressive vaccine
- Eczema vaccinatum
- Severe generalized vaccinia
- Periocular infections resulting from inadvertent inoculations

Administered IM at 0.6 mL/kg.

The dose is large, so it is given in divided doses over a 24- to 36-hour period.

May be repeated in 2–3 days if there is no improvement.

Since availability is limited, it should be reserved for the most serious cases.

➤ CONTRAINDICATIONS TO VACCINATION

- Eczema
- Malignancy
- Patients who are receiving chemotherapy, radiation therapy, or a high dose of steroids
- HIV infection
- Hereditary immune deficiency disorders
- Pregnant women
- Patients with known heart disease (cardiomyopathy, previous heart attacks, angina, or coronary artery disease)

You may give VIG at a dose of 0.3 mL/kg to persons with contraindications who have been in close contact with a smallpox patient, or to an individual who is at risk for occupational reasons.

➤ DECONTAMINATION

Standard precautions for healthcare workers.

Respiratory protection and isolation.

Soap and water decontamination.

Distinguishing Smallpox from Chicken Pox

Disease Characteristics	**Smallpox** *Variola*	**Chicken Pox** *Varicella*
Incubation	7–17 days	14–21 days
Prodromal symptoms	2–4 days	Minimal
Distribution	Centrifugal	Centripetal
Lesion development	Synchronous	Asynchronous
Scabs develop	10–14 days	4–7 days
Scabs separate	14–28 days	<14 days
Infectious	Scabs separates	Scab develops

Yersina pestis is the bacteria that causes plague. Plague is a zoonotic disease caused by rodents (rats, mice, ground squirrels). Fleas that live on the infected animal can transfer the disease by biting humans. The organism remains viable in water, moist soil, and grains for several weeks. It will remain viable for months to years at near freezing temperatures but is killed by exposure to temperatures of 55°C. The organism will remain viable for some time in dry sputum, flea feces, and buried bodies but is killed with exposure to sunlight.

Plague appears in three forms: bubonic, septicemic, and pneumonic.

➤ BUBONIC PLAGUE

Bubonic plague is transmitted to humans by the bite of an infected flea.

The incubation period of 2–10 days is followed by an acute onset of fulminant nonspecific symptoms: high fever, malaise, headache, myalgias, and sometimes nausea and vomiting. Half of the patients will experience abdominal pain.

Simultaneously or shortly after the onset of the nonspecific symptoms, patients will develop a bubo—a very painful, swollen, infected lymph node.

Buboes are most commonly seen in the femoral or inguinal lymph nodes because the legs are the most common sites for fleabites.

The liver and spleen are often palpable. One-quarter of patients will develop some form of skin lesion, pustule, vesicle, escher, or papule, in the lymphatic drainage of the bubo. This represents the site of the fleabite. Secondary septicemia is common; 80 percent of blood cultures will be positive. Only one-quarter of patients will progress to clinical septicemia.

➤ SEPTICEMIC PLAGUE

Plague septicemia can produce thromboses in the vessels of the terminal appendages (fingers, toes, penis, and nose), with necrosis, gangrene, and disseminated intravascular coagulation (DIC). Black necrotic appendages and more proximal purpuric lesions are often present.

➤ PNEUMONIC PLAGUE

Pneumonic plague is an infection of the lungs that is caused by inhalation of the organism (primary pneumonic plague) or by its spreading to the lungs during septicemia (secondary pneumonic plague). Pneumonic plague would be the predominant type of infection that is seen after an intentional release.

The incubation period is 1–6 days (average is 2–4 and is dose dependent). Onset of symptoms is acute and often fulminant.

The first signs of the illness are high fevers, chills, headache, malaise, and myalgias. These initial signs are followed 24 hours later by a cough with bloody sputum. While bloody sputum is characteristic, the sputum can also be watery and occasionally purulent.

Nausea, vomiting, diarrhea, and abdominal pain can also be present.

A cervical bubo may be present (though rarely) after an inhalation exposure.

Chest x-ray findings are variable but commonly show bilateral infiltrates that may be patchy or consolidated. The pneumonia progresses rapidly and will result in dyspnea, stridor, and cyanosis.

The disease terminates with respiratory failure and circulatory collapse.

Laboratory findings are nonspecific with a total leukocytosis of 20,000 with increasing bands and greater than 80 percent polymorphonuclear cells. There may also often be an increase in fibrin split products, indicating low-grade DIC. Kidney and liver functions may be elevated as multiorgan failure develops.

The mortality of untreated bubonic plague is approximately 60 percent and is reduced to less than 5 percent with prompt, effective therapy. The mortality of untreated pneumonic plague is nearly 100 percent and survival is unlikely if treatment is delayed beyond 18 hours.

➤ MEDICAL MANAGEMENT

Usual supportive therapy should include IV therapy and hemodynamic monitoring. Hemorrhage is uncommon with DIC, and heparin is rarely indicated. Pressor agents are not needed for endotoxic shock.

Buboes rarely require any form of local care and will usually resolve once you start systemic antibiotics. Incision and drainage of buboes present a risk of aerosol contamination; aspiration is recommended for diagnostic purposes and to relieve pain.

➤ PROPHYLAXIS

No vaccine available.

Persons who have had close contact (within 2 meters) with individuals infected with pneumonic plague, or persons who have been exposed to a plague aerosol should receive antibiotic prophylaxis for 7 days or throughout the duration of the exposure risk.

Individuals who come into contact with patients with bubonic plague do not require antibiotic prophylaxis, but you need only to observe them for 1 week for the onset of symptoms.

If patients receiving antibiotic prophylaxis become symptomatic, stop prophylaxis and begin treatment.

➤ ISOLATION AND DECONTAMINATION

Use standard precautions for bubonic plague and respiratory-droplet precautions for suspected pneumonic plague.

Soap and water are effective if decontamination is needed.

Take measures to prevent local disease cycles if fleas and rodents are present.

➤ PLAGUE TREATMENT

Postexposure Prophylaxis

Doxycycline 100 mg PO BID for 7 days.

Alternate: Ciproflaxin 500 mg PO BID for 7 days.

Treatment

Gentamicin 5 mg IV QD
or
Streptomycin 15 mg/kg IV BID.

Alternate: Doxycycline 100 mg IV × 1 mg then 100 mg PO or IV BID or Ciproflaxin 500 mg PO BID or 400 mg IV Q12 hours.

Tularemia, also known as Rabbit Fever and Deer Fly Fever, is a zoonotic disease that humans typically acquire after skin or mucous membrane come into contact with tissues or body fluids of infected animals or from the bites of infected ticks, deerflies, or mosquitoes.

The organism, *Francisella tularensis*, can remain viable for weeks in water, soil, carcasses, and hides and for years in frozen rabbit meat. It is resistant for months to temperatures of freezing and below. It is easily killed by heat and disinfectants.

➤ CLINICAL MANIFESTATIONS

Varying incubation period of 1–21 days, average 3–5 days depending on the dose. Onset is usually acute.

There are six distinct clinical forms.

➤ TYPHOIDAL TULEREMIA

5–15 percent of the naturally occurring disease.

Occurs mainly after inhalation of infectious aerosols but can occur after intradermal or GI exposure.

It manifests as fever, malaise, and weight loss, but unlike other forms of the disease, it presents without lymphadnopathy.

Pneumonia can be severe and fulminant.

Respiratory symptoms include productive or nonproductive cough, and substernal chest discomfort may also be present.

1–3 percent mortality with treated cases, 33 percent mortality if untreated.

➤ ULCEROLANDULAR TULAREMIA

75–85 percent of cases.

Often acquired through skin or mucous-membrane exposure with blood or tissue fluids of infected animals.

Characterized by fever, chills, headache, malaise, an ulcerated skin lesion, and lymphadnopathy.

The lesion is usually located where the skin contact occurred.

➤ GLANDULAR TULEREMIA

5–10 percent of all cases.

Fever and tender lymphadenopathy.

No skin ulcers.

➤ OCULOGLANDULAR TULEREMIA

1–2 percent of all cases.

Occurs when the eyes are exposed to contaminated hands or splashed with contaminated aerosols.

Unilateral, painful, purulent conjunctivitis with preauricular or cervical lymphadnopathy.

Chemosis, periorbital edema, and small nodular lesions or ulcerations of the palpebral conjunctiva.

➤ OROPHARYNGEAL TULEREMIA

Primary ulceroglandular disease confined to the throat.

Acute exudative or membranous pharyngitis with cervical lymphadnopathy.

➤ PNEUMONIC TULEREMIA

Seen in 30–80 percent of typhoidal cases and 10–15 percent of ulceroglandular cases.

Severe atypical pneumonia, fulminant with a high fatality rate if untreated.

Primary disease after inhalation of organisms.

Secondary disease after hematogenous spread.

➤ PRE-EXPOSURE PROPHYLAXIS

Pre-exposure prophylaxis with Ciproflaxin or Doxycycline used for anthrax or plague may offer protection against tuleremia.

A 2-week course of the following antibiotics given within 24 hours after aerosol exposure is effective in preventing the disease.

- Ciproflaxin 500 mg PO Q12 hours
- Doxycycline 100 mg PO Q12 hours
- Tetracycline 500 mg PO Q6 hours

Chemoprophylaxis is not recommended following a natural exposure.

➤ TREATMENT

Streptomycin 15 mg/kg IV BID
or
Gentamicin 5 mg/kg IV QD for 10 days.

Alternate: Doxycycline 100 mg IV or PO BID × 14–21 days
or
Ciproflaxin 400 mg IV (or 750 mg PO) BID for 14–21 days.

➤ ISOLATION AND DECONTAMINATION

Standard precautions.

Organisms are killed by mild heat and standard disinfectants.

All viral hemorrhagic fevers (VHF) with the exception of dengue can be transmitted to humans by aerosol inhalation.

VHFs are a diverse group of illnesses caused by the RNA of four viral families.

➤ *ARENAVIRIDAE* FAMILY

Argentine

Bolivian

Brazilian

Venezuelan

All these viruses are endemic in these countries.

Transmitted to humans by the inhalation of dust contaminated with rodent excreta.

➤ *BUNYAVIRIDAE*

Congo–Crimean Hemorrhagic Fever (CCHF).

- Tick-borne disease

Rift fever.

- Mosquito-borne disease

Hanta Virus.

- Rodent-borne disease

➤ *FILOVIRIDAE*

Ebola and Marburg.

Spread by direct contact with infected blood, secretions, organs, and semen.

➤ *FLAVIVIRIDAE*

Yellow fever and dengue.

- Mosquito-borne disease

➤ CLINICAL MANIFESTATIONS

The vascular bed is the target organ.

Microvascular damage with increased vascular permeability and capillary leakage is the dominant clinical feature.

Symptoms include fever, myalgia, and prostration.

➤ PHYSICAL EXAMINATION

Conjunctival injection.

Mild hypotension.

Flushing.

Petechial hemorrhage.

Full-blown VHF will progress to shock and generalized hemorrhage of the mucous membrane.

Pulmonary, hematopoietic, and neurological involvement will occur.

➤ DIAGNOSIS

A high index of suspicion and obtaining a detailed travel history is essential to make the diagnosis of a VHF.

A large number of patients who are from the same geographic area and who present with symptoms of VHF that have developed over a short period should lead you to suspect either a natural outbreak in an endemic setting or an intentional release.

Suspect VHF in any patient who presents with a severe febrile illness and associated vascular involvement and who has traveled to a VHF-endemic country or if intelligence suggests that the potential for an attack exists.

Thrombocytopenia and leukopenia are the rule for VHFs. The exception is that you will not see thrombocytopenia with Lassa and you will not see leukopenia with Lassa, Hantaan, and severe CCHF.

Make a definitive diagnosis by specific viral serology. With the exception of dengue, you will require specialized microbiologic containment for the safe handling of these viruses. You should consult the Centers for Disease Control and/or the US Army Medical Research Institute of Infectious Diseases for assistance with a suspected case of VHF.

➤ MEDICAL MANAGEMENT

General principles of supportive therapy apply to hemodynamic, hematologic, pulmonary, and neurologic manifestations of the disease.

Only intensive care will save the most severely ill patients.

➤ PROPHYLAXIS

Currently, the only licensed vaccine available is for yellow fever.

➤ ISOLATION AND CONTAINMENT

VHF patients generally have high levels of the virus in their blood and other secretions and are prone to aerosol nosocomial spread.

Documented cases of VHF have been reported in contacts without parenteral exposure (needle stick).

Strict standard blood- and body-fluid precautions will prevent nosocomial transmission of VHFs.

Hospitalize the patient in a private room with an adjoining room for the donning of protective equipment, the storage of supplies, and the decontamination of laboratory samples.

All persons entering the room should wear gloves and gowns. Surgical masks and eye protection is required for all those coming within 3 feet of the patient.

Nonrecirculated negative pressure rooms are recommended for patients with cough, hemorrhage, or diarrhea.

Upgrade respiratory protection to the use of a fit-tested HEPA-filtered respirator, a battery-powered air-purifying respirator, or a positive-pressure respirator if patients have prominent cough, vomiting, or diarrhea.

Section III: Toxins

Ricin is a potent cytotoxin derived from the castor bean plant.

Ricin can be produced relatively easily, inexpensively, in large quantities, and with minimal technology.

Ricin can be prepared in a liquid or crystalline form, or it can be made into a dry powder. It can be disseminated as an aerosol, injected into a target, or used to contaminate food or water on a small scale.

The clinical picture of an intoxicated victim will depend on the route of exposure. After aerosol exposure, signs and symptoms will depend on the dose inhaled.

➤ CLINICAL MANIFESTATIONS

Lethal human exposures have not been described, but the severe pathologic changes seen in animal models, such as respiratory necrosis and severe respiratory flooding, are consistent with adult respiratory distress syndrome (ARDS) and respiratory failure.

Time to death is dose dependent, usually 36–72 hours after exposure.

After inhalation, humans would be expected to develop severe lung inflammation with progressive cough, dyspnea, cyanosis, and pulmonary edema.

Sublethal exposures are characterized by:

- acute onset within 4–8 hours
- fever
- chest tightness and cough

- dyspnea
- nausea
- arthralgias

The onset of profuse sweating some hours after the onset of symptoms is commonly the sign of the termination of most of the symptoms.

By other routes of exposure, ricin is not a pulmonary irritant, but intravascular injection can cause minimal pulmonary edema.

Ingestion causes necrosis of the gastrointestinal epithelium, local bleeding, and hepatic, splenic, and renal necrosis.

Injection causes severe local necrosis of muscle and regional lymph nodes, with moderate visceral organ involvement.

➤ DIAGNOSIS

Acute lung injury affecting a large number of geographically clustered cases should raise the index of suspicion that a pulmonary irritant, such as ricin, is present.

Other pulmonary agents can present with similar signs and symptoms:

- SEB
- Q fever
- tularemia
- plague
- phosgene

➤ MANAGEMENT

Aggressive respiratory support is indicated for patients with pulmonary intoxication. Treat pulmonary edema as indicated.

You can best manage GI intoxication by vigorous gastric lavage followed by the use of cathartics. Activated charcoal is of little value because ricin molecules are too large to be absorbed. Volume replacement of GI fluid losses is important.

Provide supportive care for percutaneous exposures.

➤ DECONTAMINATION

The protective mask is effective against aerosol exposures.

Standard precautions for healthcare workers.

Ricin is nonvolatile, and secondary aerosols are not a danger to healthcare workers.

Decontamination with soap and water will inactivate ricin.

The botulinum toxins are a group of seven related neurotoxins produced by the spore-forming bacillus, *Clostridium botulinum*. These toxins, types A–G, are the most potent neurotoxins known to man.

Botulism can be delivered by aerosol or used to contaminate food or water supplies. When inhaled, these toxins produce a clinical picture similar to food-borne intoxication, although the time of onset of paralytic symptoms may be longer than for food-borne cases. Paralytic symptoms may vary according to the type and dose of toxin.

The toxins are detoxified in air within 12 hours. Sunlight inactivates the toxins within 1–3 hours. Heat will destroy toxins in 30 minutes at 80°C and in several minutes at 100°C. Water treated to 3 mg/L free available chlorine (FAC) will inactivate 99.7 percent of the toxin. Water treated to 0.4 mg/L FAC, as is done during municipal water treatment, will deactivate 84 percent of the toxin.

➤ MECHANISM OF TOXICITY

Prevents the release of acetylcholine blocking the transmission at the neuromuscular junction. This presynaptic inhibition affects both cholinergic autonomic (muscarinic) and motor (nicotinic) receptors. This interruption causes cranial-nerve and skeletal-muscle paralysis.

➤ CLINICAL MANIFESTATIONS

Onset of symptoms occurs from 12–36 hours following exposure but can vary depending on the amount of the toxin absorbed.

Cranial-nerve palsies occur early, with blurred vision that is usually due to:

- mydriasis
- diplopia
- ptosis and photophobia

Other cranial-nerve palsies are also seen:

- dysarthria
- dysphonia
- dysphagia

Flaccid skeletal-muscle paralysis progresses in a symmetric, *descending*, and progressive manner.

Respiratory failure will occur as the descending motor weakness involves the diaphragm and accessory respiratory muscles. Respiratory failure may occur abruptly.

Progression of respiratory failure, from the onset of symptoms, has occurred in as little as 24 hours in cases of severe food-borne botulism.

The autonomic effects are manifested by typical anticholinergic signs:

- dry mouth
- ileus
- constipation

- urinary retention
- nausea and vomiting may occur as nonspecific symptoms due to the ileus

Mydriasis is seen in 50 percent of cases.

Botulism toxins do not cross the blood–brain barrier, so you will not see CNS symptoms. You will also not see sensory symptoms.

➤ PHYSICAL EXAMINATION

Patient will be alert, oriented, and afebrile.

Postural hypotension may be present.

Mucous membranes will be dry and crusted.

Patient may complain of dry mouth, sore throat, and difficulty in speaking and swallowing.

Gag reflex may be absent.

Pupils will be dilated and may be fixed.

Ptosis and extraocular muscle palsies may occur.

Variable degrees of skeletal-muscle weakness may occur.

Deep tendon reflexes will be absent or present.

Cyanosis may be present with respiratory muscle paralysis.

➤ DIAGNOSIS

The occurrence of an epidemic of afebrile patients with progressive symmetric descending flaccid paralysis would strongly suggest botulism intoxication.

The Tensilon® test may be transiently positive in botulism, so it is not reliable for distinguishing botulism from myasthenia.

The cerebral spinal fluid in botulism is normal. Mental status changes seen with encephalitis are not seen with botulism intoxication.

It may become necessary to distinguish the anticholinergic symptoms of botulism from the cholinergic symptoms of nerve-agent exposure.

You can distinguish an atropine overdose from botulism intoxication by its CNS excitation (hallucinations and delirium), even though the mucous membranes are dry and mydriasis is present.

➤ MEDICAL MANAGEMENT

Airway support.

Antitoxin:

- available from the CDC for food-borne botulism
- has all the disadvantages of horse-serum products including anaphylaxis and serum sickness
- use of the antitoxin requires skin testing for horse-serum sensitivity prior to administration of the antitoxin

➤ DECONTAMINATION

Standard precautions for healthcare workers.

Toxin is not dermally active and aerosols are not a hazard to patients.

Decontaminate with soap and water.

Staphylococcal Enterotoxin B (SEB) is an enterotoxin produced by *Staphylococcus aureus*. SEB is one of the pyogenic toxins that commonly cause food poisoning in humans.

SEB is the second most common cause of food-borne illnesses. While the SEB toxin will not cause significant morbidity, it can render 80 percent of exposed individuals clinically ill. The toxin produces different clinical syndromes depending on whether it is inhaled or ingested.

➤ CLINICAL MANIFESTATIONS

Symptoms of SEB begin 3–12 hours after inhalation or 4–10 hours after ingestion. Symptoms include nonspecific flulike symptoms and specific symptoms dependent on the route of exposure.

You will see GI symptoms, nausea/vomiting/diarrhea, with oral exposure.

SEB inhalation produces predominantly respiratory symptoms: nonproductive cough, retrosternal chest pain, and dyspnea.

GI symptoms may accompany respiratory exposure due to swallowing of the agent.

Fever may last for up to 5 days and be as high as 103–106°F, with variable degrees of chills and prostration. Cough may linger for up to 4 weeks.

Physical examination is often unremarkable:

- conjunctival injection may be present
- postural hypotension may occur due to fluid losses
- chest examination is unremarkable with the exception of the rare pulmonary edema
- chest x-ray is generally normal

Diagnosis of SEB is based on clinical and epidemiologic features.

Naturally occurring staphylococcal food poisoning will not present with pulmonary symptoms.

Medical management is supportive.

➤ DECONTAMINATION

Standard precautions for healthcare workers.

SEB is not dermally active. There is no risk of aerosol exposure from contaminated patients.

Decontaminate with soap and water.

Trichothecene (T-2) mycotoxins are a group of 40 compounds produced by a common grain mold. Previously described as "Yellow Rain," they are dermally active and cause blisters after a relatively short exposure. With T-2 mycotoxins, you can expect dermal, ocular, respiratory, and gastrointestinal exposures.

➤ MECHANISM OF TOXICITY

Poorly understood.

Markedly cytotoxic to rapidly dividing cells, bone marrow, GI tract, skin, and germ cells.

Imitates the effects that radiation exposure would have on the hematopoietic and lymphoid tissue.

Mycotoxins are referred to as "radiomimetic agents."

➤ CLINICAL FEATURES

T-2 mycotoxins can adhere to and penetrate the skin, or can be inhaled or ingested.

Symptoms from all four forms of exposure can coexist.

Contaminated clothing can serve as a reservoir for further toxin exposure.

➤ DERMAL EXPOSURE

Early symptoms begin within minutes after exposure.

Burning skin pain.

Redness, tenderness, blistering.

Progression to skin necrosis with leathery blackening and sloughing off of large areas of skin.

➤ UPPER RESPIRATORY EXPOSURE

Nasal itching, pain, sneezing, epistaxis, and rhinorrhea.

➤ PULMONARY EXPOSURE

Dyspnea, wheezing, and cough.

➤ MOUTH AND THROAT EXPOSURE

Pain with blood-tinged sputum and saliva.

➤ GI EXPOSURE

Anorexia and nausea.

Vomiting and watery or bloody diarrhea.

Crampy abdominal pain.

➤ OCULAR EXPOSURE

Pain.

Tearing, redness, foreign-body sensation.

Blurred vision.

➤ SYSTEMIC TOXICITY

Can occur via any route of exposure.

Weakness, prostration, dizziness, ataxia, loss of coordination.

Tachycardia, hypothermia, and hypotension may follow in fatal cases.

➤ DIAGNOSIS

Clinical and epidemiologic findings provide clues to the diagnosis.

Finding dead animals of multiple species and physical evidence such as yellow, red, green, or other pigmented oily liquids on the ground or surrounding vegetation may suggest mycotoxins.

➤ MEDICAL MANAGEMENT

No specific therapy, and management is supportive.

Standard burn care for skin involvement.

Activated charcoal for ingested mycotoxins.

Irrigation for the eyes.

➤ DECONTAMINATION

Remove outer garments within 4 hours after exposure.

Thoroughly wash skin with soap and water.

Secondary aerosols are not a hazard, but contact with contaminated clothes and skin can produce secondary dermal exposures.

Contact precautions are indicated until decontamination is completed.

Section IV: Chemical Agents

Substances that induce a biological effect by binding to acetylcholinesterase (Ache), causing an accumulation of acetylcholine at the neuromuscular junction.

The physical signs of nerve-agent exposure are due to excess acetylcholine leading to a "cholinergic crisis."

➤ CLASSIFICATION OF NERVE AGENTS

G Agents:

GB: sarin, GA: tabun, GD: soman

- Nonpersistent
- Volatile (sarin is the most volatile)
- Vapor hazard is greater than liquid hazard
- Vapor is five to six times heavier than air and will settle in low-lying areas
- Essentially odorless and tasteless

VX:

- Liquid consistency of motor oil
- Persistent
- Low volatility
- Primarily a liquid hazard
- Evacuation to higher elevations or upper floors would be appropriate in place locations

➤ BASIC CHARACTERISTICS OF NERVE AGENTS

- Odorless
- Tasteless
- Colorless

Effects may be delayed after liquid exposure.

➤ RECEPTORS AFFECTED

Muscarinic

1. Smooth muscles
   - miosis
   - bronchoconstriction
   - increased peristalsis (stimulation of the gastrointestinal (GI) tract)
   - vasodilatation: hypotension—terminal event
   - bradycardia
2. Exocrine glands
   - bronchorrhea
   - rhinorrhea
   - salivation
   - lacrimation
   - sweating
3. Cranial nerves

Nicotinic

Skeletal Muscles

Initially local or generalized muscle fasciculations, depending on the type of exposure (dermal or vapor), followed by:

- twitching
- muscle fatigue
- flaccid paralysis
- convulsions

Tachycardia and hypertension are signs of preganglionic stimulation but are not reliable indicators of nerve-agent exposure.

➤ ROUTES OF EXPOSURE

Inhalation (Vapor Exposure)

Rapid onset seconds to minutes after exposure and is dose dependent.

No delay in onset as with liquid exposure.

Exposure is classified as small, moderate, or large.

- *Small Vapor Exposure (Local Effects)*
  Miosis, rhinorrhea, slight bronchoconstriction and dyspnea, increased secretions
- *Moderate Vapor Exposure (Local Effects)*
  Miosis, rhinorrhea, slight bronchoconstriction, increased secretions, moderate to marked dyspnea
- *Large Vapor Exposure*
  Same as for moderate exposure plus:
  - loss of consciousness
  - seizures
  - generalized fasciculations
  - flaccid paralysis
  - apnea
  - death

Liquid (Dermal Exposure)

Onset may be delayed from minutes to hours and can be as long as 18 hours. Rapid onset of symptoms can be seen with large exposures. Symptoms can develop after decontamination due to the depot effect of the agent.

Onset and penetration is influenced by site of exposure, skin temperature, and moisture.

Miosis is not an indicator of liquid exposure.

Exposure is classified as mild, moderate, or severe.

- *Mild Liquid Exposure*
  Asymptomatic period up to 18 hours
  Localized sweating and muscle fasciculations at the site of exposure
- *Moderate Liquid Exposure*
  Asymptomatic period up to 18 hours
  Symptoms same as for mild exposure, plus:
  - nausea
  - diarrhea
  - generalized weakness
  - generalized abdominal cramping
- *Severe Liquid Exposure*
  Onset 30 minutes to 2 hours
  Symptoms same as for moderate exposure, plus:
  - loss of consciousness
  - seizures
  - generalized fasciculations/flaccid paralysis
  - apnea
  - generalized secretions

➤ MANAGEMENT OF NERVE-AGENT EXPOSURE

Protect yourself.

Terminate the exposure.

Remove patient from agent or agent from the patient.

High airway resistance may be present due to bronchoconstriction and increased secretions before atropine therapy is initiated. Positive-pressure ventilation or bag-valve-mask ventilations may be required for extended periods of time to maintain adequate ventilations.

➤ NERVE-AGENT ANTIDOTES

Nerve Agent Antidote Kits (NAAKs), previously called Mark I Kits, were earlier only available to the military. NAAKs are now available to civilian agencies for the treatment of nerve-agent exposures. Nerve-agent antidotes are also supplied in standard prefilled syringes. Emergency medical personnel should be familiar with the various forms in which nerve-agent antidotes are supplied.

Each NAAK contains two auto-injectors: an Atropen containing 2 mg of atropine and a ComboPen containing 600 mg of 2-Pam Chloride. Each auto-injector is administered IM.

Additionally, "Cana" auto-injectors containing 10 mg of diazepam are available to prevent and treat seizures associated with nerve-agent exposures.

Atropine

Cholinergic-blocking drug—*muscarnic only.*

Dries secretions, reduces smooth muscle contractions.

Has no effect on skeletal muscles.

End Points of Treatment

Reduction in oral, nasal, and pulmonary secretions.

Improved shortness of breath.

Reduced airway compliance.

Important Treatment Decisions

Do not titrate treatment to

- heart rate (this is a variable sign)
- miosis (can persist for up to 6 weeks)
- twitching or fasciculations (nicotinic stimulation)

Complications of Atropine Therapy

Symptoms consistent with anticholinergic crisis.

Dry mouth, blurred vision, urinary retention, tachycardia, and the inability to sweat.

Symptoms are considered minor and can last for 24–48 hours.

Atropine is available in 2-mg auto-injectors (NAAK) and 2-mg prefilled syringes.

2-Pam Chloride

Affects nicotinic receptors.

Increases skeletal-muscle strength.

Has no effect on muscarinic receptors.

Breaks the bond between the nerve agent and Ache.

Effective unless aging has occurred.

(Aging occurs when the nerve agent irreversibly binds to Ache.)

Aging occurs in 2 minutes with exposure to soman, 3–4 hours with exposure to sarin.

Side Effects

Blood pressure elevation is an adverse effect of high dosages or rapid administration of medication.

The ComboPen that is part of the NAAK contains 600 mg of 2-Pam Chloride.

The current recommended dose of 2-Pam Chloride is 30 mg/kg or a maximum single dose of 2 g.

Diazepam (Cana)

Anticonvulsant.

10-mg auto-injector.

Used for:

- seizures
- symptoms of nerve-agent exposure that occur in two or more body systems

Effective anticonvulsant dose 30–40 mg.

Pyridostigmine Bromide (PB)

Used as a pretreatment against the rapid aging effects of soman.

Creates a reserve of free acetylcholinesterase by binding to the Ache before the nerve-agent exposure.

Protects 20–40 percent of Ache, enough to maintain spontaneous respirations after exposure to soman.

The 2-Pam Chloride frees the PB from the Ache, allowing it to metabolize the excess acetylcholine.

Dose

30 mg tablet Q8 hours.

Protected within 3–3.5 hours after first dose.

At risk, 3 hours after stopping.

➤ TREATMENT RECOMMENDATIONS BASED ON SYMPTOMS

Treatment should be based on the signs and symptoms caused by the particular exposure.

Suspected Exposure/Asymptomatic

Decontaminate the patient and observe for 18 hours.

Vapor Exposure

- *Minimal Vapor Exposure*
  Miosis with or without rhinorrhea, nausea and vomiting
  Less than 5 minutes of exposure: 1 NAAK
  Greater than 5 minutes of exposure: observe
- *Mild Vapor Exposure*
  Miosis, rhinorrhea, mild dyspnea, nausea and vomiting
  Less than 5 minutes of exposure: 2 NAAKS
  Greater than 5 minutes of exposure: 0 or 1 NAAK, depending on the severity of the dyspnea
- *Moderate Vapor Exposure*
  Miosis, rhinorrhea, moderate to severe dyspnea, nausea and vomiting
  Less than 5 minutes of exposure: 3 NAAKS and diazepam
  Greater than 5 minutes of exposure: 1–2 NAAKS

- *Moderately Severe Vapor Exposure*
  Severe dyspnea, GI or neuromuscular signs
  3 NAAKS, diazepam, and respiratory support
- *Severe Vapor Exposure*
  Loss of consciousness, seizures, flaccid paralysis, apnea
  3 NAAKS, diazepam, respiratory support, and suction

Dermal Exposure

- *Mild Dermal Exposure*
  Local sweating, fasciculations
  1 NAAK
- *Moderate Dermal Exposure*
  GI signs and symptoms
  1 NAAK
- *Moderately Severe Dermal Exposure*
  GI and neuromuscular signs and symptoms and respiratory compromise
  3 NAAKS and respiratory support
- *Severe Dermal Exposure*
  Symptoms are the same as for severe vapor exposure
  3 NAAKS, diazepam, and respiratory support

Vapor Exposure Nerve Agent Treatment Table

Vapor Exposure	Symptoms	Treatment
Minimal	Miosis with or without rhinorrhea	<5 min 1 NAAK >5 min observe
Mild	Miosis, rhinorrhea, N/V, mild dyspnea	<5 min 2 NAAKS >5 min 0–1 NAAK
Moderate	Miosis, rhinorrhea, N/V moderate to severe dyspnea	<5 min 3NAAKS + Diazepam >5 min 1–2 NAAKS
Moderately severe	Severe dyspnea, GI or neuromuscular symptoms	3 NAAKS, Diazepam, respiratory support
Severe	Loss of consciousness, apnea seizures, flaccid paralysis	3 NAAKS, Diazepam, respiratory support

Dermal Exposure Nerve Agent Treatment Table

Dermal Exposure	Symptoms	Treatment
Mild	Localized sweating fasciculations	1 NAAK
Moderate	GI sign and symptoms	1 NAAK
Moderately severe	GI and neuromuscular signs and symptoms, respiratory compromise	3 NAAKS respiratory support
Severe	Symptoms are the same for severe vapor exposure	3 NAAKS, Diazepam respiratory support

➤ GENERAL PRINCIPLES

Patients will present in large numbers soon after an outbreak of a nerve agent and may continue to present for several days after the outbreak.

Have a high index of suspicion for mixed agents (sarin and mustard gas) and secondary injuries (blasts, trampling).

Most patients and their parents who present for care will:

- be mildly contaminated and ambulatory
- not have been exposed to the agent, "walking worried"
- be psychologically traumatized
- not have been decontaminated

Prepare for patient behavioral outbursts, child-care and security issues, and the media.

Prevent cross contamination of hospital staff.

Inform the appropriate nonmedical authorities of the outbreak of nerve agent.

➤ DIFFERENTIAL DIAGNOSIS

- *Nerve Agent/Pesticide*
  Hypoxia
  Miosis
  Profuse secretions
- *Cyanide*
  Unconscious
  Metabolic acidosis
  Nonhypoxemia

- *Progressive Respiratory Symptoms*
  Consider botulism, phosgene, anthrax, and plague

➤ CLINICAL MANIFESTATIONS

Children may exhibit only central nervous system (CNS) effects from nerve-agent exposure.

- *Neuromuscular*
  Twitching, weakness, paralysis, respiratory failure
- *Effects on the Autonomic Nervous System*
  Reduced vision and miosis, drooling, sweating, diarrhea, nausea, abdominal pain, vomiting
- *Effects on the Central Nervous System*
  Headache, convulsions, coma, respiratory arrest, confusion, slurred speech, respiratory depression

Persistent miosis indicates a significant exposure.

RBC cholinesterase levels are not helpful in a mass-patient situation.

➤ TREATMENT

Base treatment on clinical suspicion.

ABCs.

Terminate exposure.

Triage infants and children as immediate due to higher susceptibilities and difficult airways.

Decontaminate:

- remove patients' clothes, then bag and seal the clothes
- wash patients with copious amounts of soap and water, then rinse
- consider dry decontamination if needed
- observe for 18 hours if the possibility exists for liquid exposure to skin or mucous membranes

Base the administration of nerve-agent antidotes on presenting symptoms.

Asymptomatic

Treatment: observe

Triage category: delayed

Constricted Pupils or Mild Rhinorrhea

Treatment: observe or admit to hospital

Triage category: delayed

Constricted Pupils and Other Symptoms of Cholinergic Crisis

Treatment:

Atropine 0.05 mg/kg IV/IM/IO to a maximum of 4 mg.

Repeat as needed every 5–10 minutes until pulmonary resistance improves or secretions resolve.

2-Pam Chloride 25–50 mg/kg IV/IM to 2000 mg maximum per hour.

Repeat every one hour as needed.

Observe for muscle rigidity, laryngospasm, tachycardia, and hypertension.

Seizures/Neurological effects

Midazolam 0.15–0.2 mg/kg IM/IV. Repeat as necessary or start continuous IV

Diazepam: rectally or IV

30 days to 5 years: 0.05–0.3 mg/kg IV

Maximum dose of 5 mg/dose

5 years and older: 0.05–0.3 mg/kg IV

Maximum dose of 10 mg/dose

Lorazepam 0.05–0.1 mg/kg IV or IM

Triage category: immediate

Apnea, Convulsions, Cardiopulmonary Arrest, or Rapid Progression of Symptoms

Treatment:

Atropine 0.05–0.10 mg/kg IV/IM/IO.

Repeat every 5–10 minutes with no maximum dose.

If using an endotracheal tube, increase the dose two to three times, mixing with 3–5 cc of normal saline.

2-Pam Chloride 25–50 mg/kg IV or IM with a maximum dose of 2000 mg/hour.

Seizures

As noted above.

Triage category: immediate

Use a nondepolarizing agent if rapid sequence intubation (RSI) is needed for intubation.

Supportive Care

Protect the airway.

Oxygen, bronchodilators, nasogastric intubation.

Cardiac monitor.

Fluids, monitor electrolytes.

- Nursing mothers should discard breast milk

Prevent hypothermia and hyperthermia.

Eye care:

- consider treating eye pain and miosis
- atropine will not reverse miosis

Treat complicating injuries and infections.

Mustard is an oily liquid with a color ranging from light yellow to brown and with an odor similar to that of garlic, onion, or mustard. Do not rely on the odor for detection.

Mustard vapor and liquid can readily penetrate the thin layers of fabrics to reach the underlying skin. Penetration is rapid and is enhanced by moisture, heat, and thin skin.

Mustard binds irreversibly to tissues within several minutes after contact. If decontamination is not done immediately after exposure, there is no way to prevent injury, although later decontamination may prevent a more severe lesion.

The clinical effects of mustard are delayed. Signs and symptoms may appear as early as 2 hours with a high-dose exposure and as long as 48 hours with a low-dose exposure. The typical time of onset is 4–8 hours. The concentration of the mustard vapor, time of exposure, ambient weather, and body site exposed are factors in the time of onset.

Mustard causes tissue damage within several minutes after contact with the skin without causing any clinical signs. This causes the contaminated person to be unaware that he is exposed. To prevent injury, carry out the decontamination immediately after contact.

➤ CLINICAL MANIFESTATIONS

Skin

Erythema is the earliest and mildest indication of skin involvement after mustard exposure. The erythema begins 2–24 hours after vapor exposure and is dependent on ambient temperature, humidity, and the skin site exposed.

The thin moist areas of skin such as axilla, perineum, external genitalia, antecubital fossa, and neck are the most sensitive to mustard exposure.

Small vesicles that will later coalesce to form bullae will develop within the erythematous areas. The typical blister is large, dome shaped, thin walled, and translucent. *The fluid does not contain mustard and is not a danger to medical staff.*

Treatment

Treat erythema with a soothing lotion or cream.

Leave small blisters intact, aspirate large blisters, leaving skin intact.

Irrigate denuded areas daily with normal saline or clean soapy water.

Cover these areas with antibiotic ointment or silvadene cream.

Prescribe oral analgesics for pain control.

Monitor fluids and electrolytes, but fluid losses are not the same as seen with thermal burns. Resist the temptation to overhydrate.

Pulmonary

Early effects:

- irritation or burning of the nares
- epistaxis, sinus pain, or irritation
- irritation and soreness of the pharynx

As the exposure increases, the patient will begin to experience laryngitis, voice changes, and a nonproductive cough.

With involvement of the trachea and upper airways, a productive cough will develop. As the lower airways become involved, the cough will worsen, become productive, and dyspnea will develop.

Necrosis of the upper airway mucosa with inflammation can cause pseudomembrane formation.

Respiratory failure is the most common cause of death. Mechanical obstruction by pseudomembranes and agent-induced laryngospasm are potential causes of death within the first 24 hours after exposure. Death on day 3–6 postexposure is due to bacterial pneumonia.

Treatment

Upper airway symptoms are best treated by steam inhalation and cough suppressants.

The productive cough, dyspnea, fever, and leukocytosis that develop within the first 12–24 hours postexposure is due to a sterile bronchitis or pneumonitis and does not require antibiotics.

Infection often occurs on the third day after exposure.

Intubate early before laryngeal spasm or edema develop, making airway management difficult.

Assist ventilations or provide supplemental oxygen.

Provide bronchodilators or steroids for bronchospasm.

Gastrointestinal

GI mucosa are very susceptible to mustard, either from systemic absorption or ingestion.

Nausea with or without vomiting lasts 24 hours or less.

Nausea and vomiting is not due to the direct effect of mustard on the GI tract but is a nonspecific reaction to the odor or cholinergic stimulation.

Prolonged vomiting or increased diarrhea that begin days after exposure suggest direct GI involvement by severe systemic poisoning, which is a poor prognostic sign.

Treatment

Anticholinergic drugs (atropine 0.4–0.6 mg IV or IM).

Antiemetics.

Fluids for dehydration.

Ocular

The eyes are the organs most sensitive to mustard vapor. The latent period is shorter for eye injury than for skin injury.

Irritation is the only symptom after a low-dose exposure; reddening of the eyes may be the only effect.

As the dose increases, the spectrum of injury includes progressively more severe conjunctivitis, photophobia, blepharospasm, pain, and corneal damage.

The most severe damage is caused by exposure to liquid mustard from airborne droplets or by self-contamination.

Miosis noted after mustard exposure is probably due to the cholinergic activity of mustard.

Treatment

Thoroughly irrigate the eyes.

Treat eye irritation with ophthalmic solutions (liquid tears, antibiotic ointment to prevent infection).

Instill homotropine to prevent synechiae formation.

Apply Vaseline or a similar substance to the eyelids to prevent them from sticking together and increasing the risk of scarring and forming adhesions.

Use topical analgesics initially if blepharospasm is too severe to permit adequate examination.

Avoid long-term use of topical analgesics and manage pain with oral analgesics.

Topical steroids have not been shown to be a benefit, but use in the first 48 hours postexposure may reduce inflammation.

Sunglasses may reduce photophobia.

Central Nervous System

CNS effects are poorly defined.

Seizures are possible.

Reports on people exposed to small amounts of mustard during WWI and in Iran stated that they appeared sluggish, apathetic, and lethargic.

➤ GENERAL CARE

Liberal use of analgesics and antipruritics.

Maintenance of fluid and electrolyte balance.

Parenteral food supplements and vitamins may be helpful.

➤ TRIAGE

Delayed

Triage most patients exposed to mustard as delayed.

Mild to moderate pulmonary effects will require further care but are not in the immediate category for triage.

Triage eye injuries as delayed.

Immediate

Triage patients with moderately severe to severe pulmonary signs and symptoms as immediate. Consider two factors in making this decision:

1. Patients who develop severe pulmonary symptoms within 4–6 hours of exposure will probably not survive despite maximal medical care. In this situation, it may be better to render limited medical care to other patients.
2. Evacuation to a tertiary medical center is required. The patient may survive the trip, but the delay may cause the lesion to progress to an irreversible stage.

Minimal

Triage as minimal those patients with lesions over a small body surface, less than 5 percent BSA, when the lesions are not in a vital area.

Treat and discharge patients with minor eye injuries that include irritation and reddening.

Give symptomatic treatment to and discharge patients who develop the slight respiratory symptoms of a non-productive cough and throat irritation 12 hours or longer after exposure.

Expectant

Triage as expectant those patients who develop severe pulmonary effects within 4–6 hours after exposure, with greater than 50 percent BSA burns from mustard.

Lewisite is a vesicant that causes damage to the eyes, skin, and airways by direct contact. It causes an increase in capillary permeability that produces hypovolemia, shock, and organ damage.

Lewisite is an oily, colorless liquid with the odor of geraniums. It is more volatile than mustard gas.

Lewisite vapor or liquid causes *immediate* pain and burning on contact with the skin.

➤ CLINICAL MANIFESTATIONS

The vapor of lewisite is so irritating that when exposed, a person will attempt to don a protective mask or leave the area immediately.

Because this warning causes people to decontaminate themselves immediately, a lesion from lewisite will probably not be as severe as a lesion from mustard gas.

Skin

Liquid lewisite will produce a grayish area of dead skin within 5 minutes of contact.

Redness and blister formation follow more rapidly than with mustard gas, although full lesion formation does not develop for 12–18 hours.

This lesion has more tissue necrosis and sloughing off than with mustard gas.

Eyes

Pain and blepharospasm occur on contact. This is followed by edema of the conjunctiva and eyelids. The eyes may be swollen shut within 1 hour.

High doses may cause corneal damage and iritis.

Liquid lewisite causes severe eye damage within minutes of contact.

Respiratory

The extreme irritation caused by lewisite to the nose and upper airways will cause a person to put on a gas mask and leave the area.

The airway mucosa is the primary target and damage progresses down the airways in a dose-dependent manner.

Pseudomembranous formation is common.

Pulmonary edema may complicate lewisite exposure.

Other

Lewisite causes an increase in the permeability of the systemic capillaries that results in an intravascular fluid loss, hypovolemia, shock, and organ congestion.

The organ congestion may lead to hepatic or renal necrosis, with more prominent GI effects that will cause nausea and vomiting.

➤ MEDICAL MANAGEMENT

Early decontamination is the only way to prevent or lessen the damage from exposure to lewisite.

British Anti-Lewisite (BAL) is an antidote for lewisite exposure. BAL is toxic and the user must follow the instructions on the package insert carefully.

BAL skin and ophthalmic ointments decrease the severity of skin and eye lesions when applied immediately after early decontamination (neither is currently manufactured).

➤ TRIAGE

Triage patients using the same criteria for patients exposed to mustard gas.

Phosgene oxime is an urticant or nettle agent that causes a corrosive type of skin and tissue lesion. It is not a true vesicant because it does not cause blisters.

There is very little information on CX.

➤ CLINICAL MANIFESTATIONS

Skin

CX liquid or vapor causes pain on contact, which is followed by skin blanching with a erythematous ring in 30 seconds, a wheal in 30 minutes, and skin necrosis later. The extreme pain may persist for days.

Eyes

Damage is similar to that caused by exposure to lewisite.

Pulmonary

Very irritating to the upper airways. Pulmonary edema develops after inhalation and after skin exposure.

Other

Hemorrhagic and inflammatory changes in the GI tract.

➤ MEDICAL MANAGEMENT

Supportive.

Usual skin care, same as for a necrotic lesion from any other cause.

➤ TRIAGE

Delayed.

➤ PHOSGENE

Used in the production of plastics and dyes.

Twice as potent as chlorine.

Risk of exposure from industrial accidents and domestic terrorism.

The odor of newly mown hay is not a reliable indicator of exposure.

➤ CHLORINE

First chemical weapon of modern warfare.

Precursor to the production of hydrocarbonated plastics.

➤ PERFLUOROISOBUTYLENE (PFIB)

When organoflouride polymers (Teflon) are burned, PFIB is a by-product of this combustion.

Inhalation of this toxic combustion product produces the same effects as phosgene.

➤ OXIDES OF NITROGEN

Produced by high temperature welding or burning processes.

Small arms' rounds, artillery shells, and other military explosives produce nitrogen dioxide.

➤ HC SMOKE

Standard military white obscurant smoke that is dis-bursed in grenades and artillery shells.

Zinc chloride is the by-product of combustion that is believed to cause the majority of the toxic effects.

Produces the same type of pulmonary edema as found in exposures to other pulmonary agents.

➤ GENERAL CONSIDERATIONS

Dispersed as aerosol, gas, or vapor, with inhalation as the primary route of exposure.

Inhalation will result in varying degrees of pulmonary edema, usually after a symptom-free period.

Exertion that causes an increase in respirations will increase the intensity of exposure.

Pre-existing airway injury—from disease, previous injury, or smoking—can worsen symptoms.

Pulmonary agents will likely cause bronchospasm in patients with atopic symptoms.

Bronchospasm can reduce clearance of the agent from the lungs.

Coughing, sneezing, and mucocillary clearance can provide initial protection after exposure.

➤ PHOSGENE

An asymptomatic latent period follows exposure to phosgene. With a minimal exposure, this period can be as long as 24 hours. The duration of the latent period is shorter with larger exposures and physical activity.

After the asymptomatic period but before the onset of respiratory symptoms, the patient may begin to experience irritation of the mucous membranes, increased lacrimation, conjunctivitis, nonproductive cough, and sensation of substernal pressure.

Respiratory symptoms are characterized by shortness of breath, with or without chest tightness. Fine crackles, heard initially only with forced exhalation, are followed by the development later of coarse ronchi and rales in all lung fields.

Pulmonary edema develops.

Death is due to a combination of respiratory failure, hypoxia, and hypovolemia.

The development of pulmonary edema within 4 hours after exposure indicates a poor prognosis.

Complications are due to lung damage and respiratory infections.

➤ CHLORINE

The latent period may be 1 hour or shorter.

Chlorine will chemically convert to hydrochloric acid when it comes into contact with the moist surfaces of the respiratory tract.

- *Mild Exposure*
  Sensation of suffocating or choking
  Ocular and nasal irritation
  Chest tightness and cough
  Exertional dyspnea

- *Moderate Exposure*
  Mild symptoms, hoarseness, and stridor
  Pulmonary edema within 2–3 hours
- *Severe Exposure*
  Severe dyspnea at rest
  Pulmonary edema within 30–60 minutes
  Copious upper airway secretions
  Sudden death from laryngospasm

➤ PFIB

Produces the same type of pulmonary edema as phosgene but is 10 times more toxic.

Shorter latent period than phosgene, 1–4 hours.

➤ HC SMOKE

Will cause dyspnea soon after exposure. The dyspnea will resolve but will recur within 24–36 hours.

HC-smoke exposure can lead to interstitial fibrosis, hemorrhagic fibrosis of the upper airways as well as pulmonary edema.

The potential exists for sudden death from laryngospasm.

- *Low Exposure*
  Dyspnea with or without x-ray or arterial-blood-gas (ABG) abnormalities
  Observe these patients for 4–6 hours postexposure

- *Moderate Exposure*
  Severe dyspnea that may show rapid improvement within the first 4–6 hours
- *Severe Exposure*
  Severe dyspnea and tachypnea within the first hour after exposure
  Chest x-ray will be initially unremarkable but will begin to show dense infiltrates representing pulmonary edema 4–6 hours after exposure

➤ NITROGEN OXIDES

Long-term exposure is associated with fibrotic changes in the lung.

Can induce bronchospasm in patients with asthma.

A toxic exposure may occur with minimal discomfort. Ignore the coughing and choking sensations.

A lethal exposure can occur after only 30 minutes, with the patient experiencing only minimal discomfort.

➤ TREATMENT OF PULMONARY AGENTS

Terminate the exposure.

ABCs as required.

Manage secretions and treat bronchospasm.

Reserve antibiotics for patients with infections documented by sputum gram stain and culture.

Prevent and treat pulmonary edema with positive end expiratory pressure (PEEP) or continuous positive airway pressure (CPAP).

Prevent and treat hypoxia.

Treat hypotension.

Encourage rest.

➤ TRIAGE GUIDELINES

- *Immediate*
  Patients who develop pulmonary edema within 12 hours of exposure
  Shorter onset of pulmonary edema signifies a worse prognosis
- *Delayed*
  Patients experiencing dyspnea without objective signs of respiratory distress
  Observe closely and retriage hourly
- *Minimal*
  An asymptomatic patient with a known exposure
  Observe and retriage every 2 hours
  If the patient remains asymptomatic for 12 hours after exposure, discharge
- *Expectant*
  A patient who presents with cyanosis, pulmonary edema, and hypotension 6 hours after exposure

➤ GENERAL CONSIDERATIONS

Use of routine prophylactic antibiotics is not indicated.

Prophylactic antibiotics encourage growth of the most resistant organisms and make subsequent treatment difficult.

Elevated temperatures, infiltrates on the chest x-ray, and elevated WBC counts are typical responses to any toxic pulmonary exposure and do not generally indicate the presence of infection within the first 3 days.

Discharge patients within 12 hours if they only experience eye or upper airway symptoms, have a normal physical examination (PE), and are asymptomatic.

Discharge patients who initially complained of dyspnea but whose chest x-rays, ABGs, and PEs are all normal after 24 hours.

Discharge patients who initially were symptomatic with abnormal chest x-rays, ABGs, or PEs but had normal PEs, ABGs, and chest x-rays after 48 hours.

A rapidly acting toxin that can cause death within several minutes of exposure. Inhaled or dermal exposures can cause death within 1 hour of contact. The agent is less rapidly absorbed after ingestion.

Produces its toxic effects by interfering with the utilization of oxygen at the cellular level, causing cellular anoxia.

Rapidly absorbed through the eyes, but inhalation is the usual route of exposure.

➤ HYDROGEN CYANIDE (AC)

Colorless, highly soluble.

Stable in water.

Faint odor of peach pits or bitter almonds.

Odor is not a reliable indicator of exposure to AC.

40 percent of the population cannot smell AC.

Vapor is lighter than air and will rise as it is released This is an important consideration when determining shelter in place locations.

➤ CYANOGEN CHLORIDE (CK)

Colorless.

Vapor is heaver than air and will settle in low-lying areas. This is an important consideration when determining shelter in place locations.

Very irritating to the eyes and mucous membranes.

Pungent, biting odor.

➤ SEVERE SYMPTOMS AFTER VAPOR EXPOSURE

Cardiac and CNS are highly susceptible.

Respiratory rate will increase approximately 15 seconds after exposure.

Seizures will develop in 15–30 seconds.

Apnea develops 2–3 minutes later.

Cardiac arrest develops within several minutes.

Total time course from exposure to death is 6–8 minutes.

➤ SYMPTOMS AFTER INGESTION OF CYANIDE

Slower onset and progression of symptoms.

Onset of effects is dependent on the amount absorbed and the rate of absorption.

Respiratory rate will increase, followed by anxiety or apprehension, agitation, vertigo, weakness, nausea with or without vomiting, and muscular trembling.

Later the patient will lose consciousness, respiratory effort will decrease in rate and effort, seizures will develop, patient becomes apenic, and arrhythmias will develop that are followed by cardiac arrest.

➤ MILD TO MODERATE EXPOSURE TO AC

Symptoms depend on the concentration of the agent and duration of the exposure.

- *Mild*
  Headache, vertigo, and nausea that persist for several hours prior to recovery

- *Moderate*
  Leg weakness, vertigo, nausea, headache occur soon after exposure and are followed by seizures and death

Residual CNS damage may persist if coma is prolonged and is manifested by:

- irrational behavior
- abnormal reflexes
- unsteady gait
- nerve deafness

➤ SYMPTOMS OF EXPOSURE TO CK

Manifested by a combination of signs produced by HC and lung irritants.

- *Mild Exposure*
  Irritation of the nose, throat, and eyes
  Increased lacrimation
- *Moderate Exposure*
  Irritation of mucous membranes
  Cough, chest tightness
- *Severe Exposure*
  Local irritation produces dyspnea
  Unconsciousness is followed by falling respirations and death within a few minutes
  Pulmonary edema may develop if not fatal

➤ LABORATORY STUDIES

Elevated blood cyanide levels.

Metabolic acidosis with high lactic acid levels; or,

Metabolic acidosis with an unexplained anion gap.

Oxygen content of venous blood will be higher than normal.

➤ PHYSICAL FINDINGS

Nonspecific.

Cherry red skin (not characteristic).

Severe respiratory distress without cyanosis.

➤ TREATMENT

Protect yourself and the patient; put on masks immediately.

Remove the source from the patient or remove the patient from the source.

Remove contaminated clothing and rinse skin with soap and water.

Dermal decontamination is only required if clothes are saturated with the agent.

For ingestions of the agent, use gastric gavage followed by administration of activated charcoal down the gavage tube.

ABCs.

Provide 100 percent oxygen or support respirations with mechanical ventilation.

Circulatory support with IV fluids and vasopressors.

Correct metabolic acidosis with sodium bicarbonate IV.

Treat seizures with benzodiazepines.

➤ ANTIDOTES

Rapidly begin infusion of sodium nitrite IV.

Adults: One to two vials in a standard cyanide kit (10 mL).

Pediatric patients: 0.33 mL/kg of 3 percent solution slow IV over 5–10 minutes.

Sodium thiosulfate

Adults: 50 mL at a rate of 3–5 mL a minute.

Pediatric patients: 0.33 mL/kg of 3 percent solution slow IV infusion over 5–10 minutes.

Second treatment at half the original dose.

Many will recover without the antidote.

Antidotes that are given too fast are also associated with patient mortality.

Titrate infusion to effects.

Signs of overdosage: nausea, vomiting, muscle cramps, hypotension, joint pain, psychosis.

➤ TRIAGE

- *Immediate*
  Patients who present within minutes of exposure with seizures or apnea but are not in cardiac arrest
  Immediate administration of the antidote will be life saving
- *Delayed*
  Recovery from mild effects or successful therapy
  Full recovery can be delayed for hours
- *Minimal*
  Inhalation less than lethal amount with mild effects
- *Expectant*
  Cardiopulmonary arrest

➤ RECOVERY

Generally, patients will die within 5–10 minutes and before reaching a definitive care facility.

A patient who reaches a definitive care facility alive has probably not received a lethal dose of cyanide.

Incapacitating agents, also known as BZ, are anticholinergic compounds similar to atropine and scopolamine.

Incapacitating agents are intended to produce temporary mental impairment in those who are exposed to the agent.

BZ is odorless, nonirritating, and may be dispersed as an aerosol or dissolved in a solvent for absorption by the skin. There is a latent period of 30 minutes to 24 hours from the time of exposure to the onset of symptoms.

BZ binds with the muscarinic receptors in cardiac and smooth muscle as well as exocrine glands. This accumulation of BZ prevents acetylcholine from binding with the receptors, decreasing the expected response from that receptor.

➤ CLINICAL MANIFESTATIONS

The clinical effects of BZ are manifested by the understimulation of the peripheral nervous system (PNS) or CNS.

Peripheral Nervous System

Effects are similar to the effects of atropine.

Skin will be hot and dry.

Inability to dissipate heat by evaporative cooling.

Body-core temperature will rise.

Dry mouth and nose.

Increased thirst.

Decreased tears.

Dilated pupils that cause visual impairment.

Urinary retention and decreased urinary force.

A mnemonic to recall the PNS effects of BZ is:

Dry as a bone
Hot as a hare
Red as a beet — **Anticholinergic Toxicity**
Blind as a bat
Mad as a hatter

Central Nervous System

Dose-dependent decrease in the level of consciousness, beginning with drowsiness and progressing through sedation to stupor and then coma.

The patient may be disoriented to time and place.

Disturbances in judgment and insight.

May exhibit inappropriate or vulgar behavior.

Patient is easily distracted and may exhibit short-term memory loss.

Despite these deficits, patients will try to make sense of their environment and make up answers on the spot to questions that confuse them.

Speech will become slurred and often senseless with a flat monotonous voice.

Handwriting deteriorates.

Patient may get undressed.

Phantom behaviors such as constant picking, plucking, or grasping motions.

Peripheral disturbances of the CNS in BZ exposure are manifested by both illusions and hallucinations. Hallucinations from BZ tend to be realistic, distinct, and easily identifiable. These effects are not only seen in individuals but also in groups of individuals exposed to BZ. Patients will share illusions or hallucinations.

➤ TIME COURSE OF EFFECTS

The clinical effects of BZ, either by inhalation or ingestion, develop after an asymptomatic period that can be from 30 minutes to 20 hours. The usual range of the asymptomatic period is 0.5–4 hours, with a mean time frame of 4 hours. Effects can appear as long as 36 hours after exposure.

The onset of symptoms after BZ exposure is dose dependent and typically lasts for 72–96 hours. The clinical effects can be divided into four stages:

1. First phase is the onset of symptoms 0–4 hours after exposure and is characterized by parasympathetic blockade and mild CNS effects.
2. Second phase is 4–20 hours after exposure and is characterized by stupor with ataxia and hyperthermia.
3. Third phase develops 20–96 hours after exposure and is characterized by full-blown delirium.

4. Fourth phase or resolution is characterized by paranoia, deep sleep, reawakening, crawling, climbing, automatisms, and eventual reorientation.

➤ MEDICAL MANAGEMENT

General supportive management includes decontamination of the skin and clothes.

You may need to physically restrain patients.

Monitor temperature for hyperthermia, especially in patients who are already dehydrated from overexertion or insufficient water intake or who are in a hot humid environment.

➤ ANTIDOTE

Physostigmine will temporarily raise acetylcholine levels by reversibly binding to Ache.

Can be used as a pretreatment.

Physostigmine is minimally effective during the first hour after exposure but very effective 4 hours after exposure.

Oral dosing requires one and a half times the dose of the antidote given IM or IV. IM injections will last only about 60 minutes, requiring frequent redosing.

When selecting a route of administration, IM or IV, be aware that if the IV infusion is too slow, nerve agent-like bradycardia can develop. An IV infusion that is given too rapidly can cause arrhythmias, excessive secretions, and seizures.

➤ DOSING

IM: 45 mcg/kg for adults (20 mcg/kg for children)

IV: 30 mcg/kg slowly (1 mg/min)

PO: 60 mcg/kg if patient is cooperative

Because of its bitter taste, mix it in juice.

Perform a mental-status examination every hour, and the dose and time interval of dosing should be titrated on the basis of improvement or decline in the patient's mental status. As the patient improves, the dose requirement will decrease.

➤ TRIAGE

- *Immediate*
  Patients with severe cardiorespiratory compromise or severe hyperthermia
- *Delayed*
  Pronounced or worsening anticholinergic signs
- *Minimal*
  Mild PNS or CNS anticholinergic effects
- *Expectant*
  Severe cardiorespiratory compromise in a situation where treatment or evacuation resources are limited

➤ DISCHARGE FROM CARE

An early discharge is probably not a realistic possibility for the majority of patients, who may require observation and management for at least several days.

Riot-control agents, also called irritants, lacrimators, and "tear gas," produce respiratory discomfort and eye irritation to temporarily incapacitate the individual.

The main effects of riot-control agents are pain, burning, and irritation of exposed mucous membranes and skin.

➤ EYES

The eye is the organ most sensitive to riot-control agents. Exposure to riot-control agents produces a sensation of ocular burning that progresses to tearing and blepharospasm. Conjunctival infection is also seen. The blepharospasm is so severe that the eyelids close tightly, producing a transient blindness.

The compounds are solids, and damage to the cornea and conjunctiva can occur if the solid material comes into contact with the eyes. Perform an eye examination for all patients who complain of eye pain.

➤ NOSE AND MOUTH

Contact with the mucous membranes of the nose and mouth will produce a burning sensation, rhinorrhea, sneezing, and increased salivation.

➤ AIRWAY

Inhalation will cause a burning sensation, productive cough, and the sensation of chest tightness or inability to breathe.

Inhalation of a riot-control agent may possibly exacerbate a chronic respiratory condition. There is no evidence that a single or repetitive exposure will cause permanent lung damage.

➤ SKIN

Contact with the skin will cause a transient tingling or burning sensation that is followed by erythema. These symptoms will be exaggerated if the skin is abraded or freshly shaven.

With extremes of temperature, humidity, and agent concentration, a more severe dermatitis may develop that will be accompanied by vesicle formation.

➤ GASTROINTESTINAL TRACT

GI effects usually do not occur, but with exposure to large concentrations of the agent, vomiting may occur.

➤ MEDICAL MANAGEMENT

Irrigate the eyes with copious amounts of water or saline. Treat conjunctivitis with antibiotic ointment.

For respiratory exposures, remove the patient to fresh air and provide supplemental oxygen or assisted ventilations if needed. Administer Albuterol by metered-dose inhaler or nebulizer for bronchospasm.

Irrigate exposed skin with copious amounts of water.

➤ TRIAGE

Field triage is usually not necessary. Patients who present with complicating injuries should be triaged according to the nature of their injuries.

Discharge patients as soon as possible.

➤ NERVE AGENTS

General-vapor exposure: miosis, increase in secretions, chest tightness.

Dermal: localized sweating and fasciculations.

➤ ANTHRAX

You will see a widened mediastinum and clear lungs on the chest x-ray.

Patient will appear toxic but will have a normal CBC/chest x-ray.

Flulike symptoms with temperatures as high as 104°F and drenching sweats.

➤ PNEUMONIC PLAGUE

Hemoptysis.

➤ BOTULISM

Symmetric descending flaccid paralysis.

Ptosis.

Normal mental status.

➤ SMALLPOX

Centrifugal distribution.

Lesions begin in hands, face, and forearms, then spread to the lower extremities.

All lesions in the same stage of development.

➤ PHOSGENE

Odor of new mown hay.

➤ CYANIDE

Cherry red lips/odor of bitter almonds.

➤ VESICANTS

Odor of garlic/oily liquid.

Mustard gas: delayed symptoms.

Lewisite: immediate symptoms.

Section V: Acute Radiation Sickness

Acute Radiation Sickness (ARS) is an acute illness caused by irradiation of the entire body (or most of the body) by a high dose of penetrating radiation, in a very short period of time.

The signs and symptoms of ARS progress through four distinct stages.

➤ PRODROMAL STAGE

The time of onset of symptoms is based on the radiation dose. How quickly patients develop vomiting and diarrhea is an indication of the amount of radiation they were exposed to. This time can vary from minutes to hours.

Onset of V&D	**Radiation dose**
Vomiting > 2 hours/no diarrhea	Mild 1–2 Gy
Vomiting 1–2 hours/no diarrhea	Moderate 2–4 Gy
Vomiting < 1 hour/mild diarrhea	Severe 4–6 Gy
Vomiting < 10 minutes/ heavy diarrhea	Lethal > 8 Gy

Transient episodes of vomiting and diarrhea lasting for 2–4 days occur with radiation doses of 1–5 Gy. Above 5 Gy, these symptoms are persistent.

Patients who are asymptomatic (no vomiting or diarrhea) at 2 hours or more after exposure have likely sustained minimal radiation exposure and survival is likely.

➤ LATENT PERIOD

Patients will become asymptomatic 2–4 days after exposure if radiation levels were between 2 and 3 Gy. This latent period can last for up to 3 weeks but can be shorter with larger radiation exposures. Irradiation of the bone marrow will cause a decrease in the total number of lymphocytes, neutrophils, and platelets during this period.

➤ ILLNESS PERIOD

Patients will develop infections and bleeding during this phase.

➤ RECOVERY OR DEATH

Recovery can take weeks to months in patients who have not received a lethal dose of radiation. Death will result from infection, bleeding, and/or shock in patients who have sustained a lethal exposure to radiation.

➤ THREE CLASSIC ARS SYNDROMES

Hematopoietic Syndrome

Earliest indicator of severe radiation exposure.

There is a rapid decline in all cell lines, leading to impaired immune response, increased risk of infections, bleeding, anemia, and impaired wound healing.

Occurs with radiation exposures between 1 and 8 Gy.

Gastrointestinal Syndrome

Manifested by nausea, vomiting, and diarrhea.

Seen with a radiation dose of greater than 8 Gy.

The onset of symptoms is directly related to the dose of radiation.

Vomiting and diarrhea lead to fluid loss and subsequent dehydration.

Intestinal mucosal cells will slough off, reducing the barrier that separates abdominal bacteria from the blood stream. Septicemia will develop.

Death can occur in 1–2 weeks.

Cardiovascular/Central Nervous System

Whole-body doses of over 29 Gy produce this syndrome.

Patients will develop vomiting and diarrhea that occur within minutes, confusion, disorientation, seizures, cerebral edema, and coma.

Recovery does not occur.

➤ LOCAL RADIATION INJURY

Most irradiated patients receive only partial-body exposures because parts of their bodies are shielded from the radiation.

Epilation

Loss of body hair occurs with a dose of 3 Gy and will develop within 7–21 days.

Erythema

Immediate erythema occurs within a few minutes of exposure and is most likely due to chemical irritation and not radiation.

Early erythema indicates a dose of approximately 6 Gy and will occasionally disappear and return later.

Take pictures to document the early erythema.

Late erythema occurs 2–3 weeks after exposure to a radiation dose of 3 Gy.

Dry Desquamation

Skin peeling with blisters occurs with a dose of 10 Gy, usually 2–4 weeks after exposure.

Wet Desquamation

Skin peeling with large or small blisters occurs with a dose of 15 Gy and can take up to 8 weeks to develop.

Large blisters require a dose of greater than 25 Gy before developing.

Necrosis

Skin necrosis requires a dose of greater than 50 Gy and can take days to months to develop.

➤ INITIAL MANAGEMENT

ABCs.

Monitor blood pressure, ABGs, electrolytes, and urine output as needed.

Treat major trauma as indicated.

Monitor fluids.

Treat vomiting to prevent dehydration.

Take a complete blood count (CBC) every 2–3 hours for the first 8–12 hours following the exposure. Continue CBCs every 4–6 hours for the following 2–3 days.

Record all clinical symptoms, particularly nausea, vomiting, diarrhea, itching, and reddening or blistering of the skin. Consult with experts in radiation-accident management.

Section VI: Contact Information and References

USAMRIID's Emergency Response line
1-888-872-7443

Domestic Preparedness Chemical/Biological
Help Line
Edgewood Operations Center
Military Use only
1-410-436-4484

National Domestic Preparedness Office
Civilian Use
1-202-324-9025

Centers for Disease Control
Emergency Response Line
1-770-488-7100

REACTS (Radiologic Emergencies)
1-865-576-1005

Acute Radiation Syndrome Fact Sheet for Physicians. CDC Publication, December 20, 2002.

Garner, A., Lee, A., Harrison, K., and Schultiz, C. Disaster medicine/domestic preparedness: comparative analysis of multiple casualty incident triage algorithms. *Annals of Emergency Medicine*, vol 38 #5, November 2001.

Guidelines for Cold Weather Mass Decontamination During a Terrorist Chemical Agent Incident. U.S. Army Soldier and Biological Chemical Command (SBCCOM), January 2002.

Guidelines for Decontamination During a Terrorist Chemical Agent Incident. U.S. Army Soldier and Biological Chemical Command (SBCCOM), January 2000.

Medical Management of Biological Casualties, Fourth Edition. U.S. Army Medical Research Institute of Infectious Diseases (USAMRIID), February 2001.

Medical Management of Chemical Casualties, Third Edition. U.S. Army Medical Research Institute of Chemical Defense (USAMRICD), July 2000.

Medical Management of Radiologic Casualties, First Edition. Armed Forces Radiobiology Institute (AFFRI), December 1999.

Rotenberg, G. and Newmark, J. Nerve agent attacks on children: diagnosis and management. *Pediatrics*, vol 112 # 3, September 2003.

Sidell, F. *Textbook of Military Medicine. Aspects of Chemical and Biological Warfare*, Chapter 5: Nerve Agents, 1997.

Simple Triage And Rapid Treatment Guidelines (START). Newport Beach Fire Department and Hoag Hospital Paramedic Department, 1994.

The Sanford Guide to Antimicrobial Therapy, 34th Edition, 2004.